My grandfather has a grand supreme truck.

Rolling and roaring down highways,
he thunders across the country.

His rig is big. It weighs a lot. It rolls on many wheels.

Its weight is eighty thousand pounds.
It is almost eighty feet long.

That weighty truck lugs freight of all kinds: tools, toys, ties, TVs!

Grandfather wields his bulky big rig smoothly on eighteen wheels!

Hither and thither he rolls along,
hauling tons of freight.

Grandfather's big rig is an eighteen wheel wonder.

Isn't that supreme truck grand?